Connecting the Dots
Mere Coincidence or Designed by a Supreme Intelligence?
You Be the Judge.

Part 1

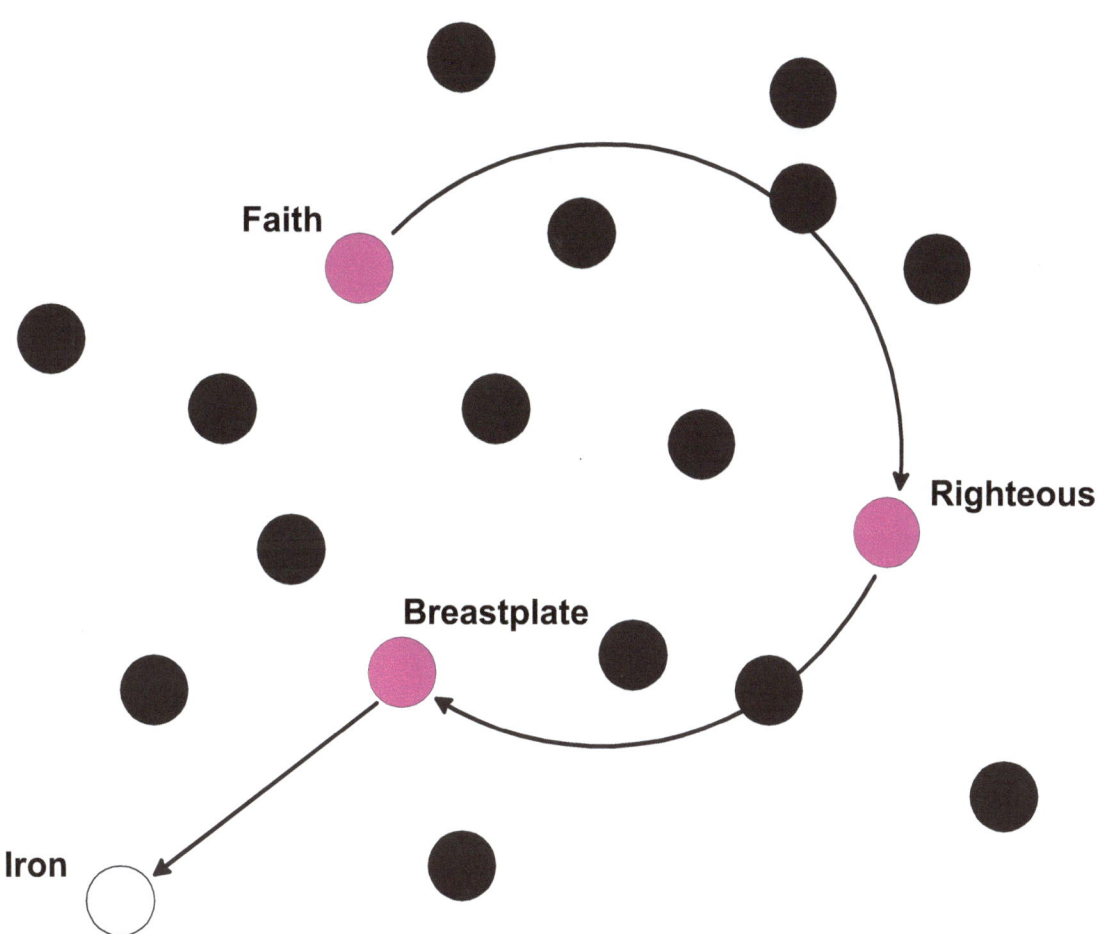

Vernon J. O'Neal

PREFACE

The purpose of the following information is not to suggest that there exists a different interpretation of the Bible's spiritual message but merely to highlight the incredible wisdom of the one who wrote it. It is hoped that this analysis will fortify the faith of some and perhaps convince many others that the greatest story ever told is unquestionably the product of a supreme intelligence.

Table of Contents

Chapter		Page
1	The Mind of the Supreme Intelligence	4
2	Connecting the Dots	6
3	Mathematical Balance	9
4	Balancing Words with Numbers	10
5	The Analysis of Boxes 7, 4, and 1	13
6	Numbers + Words = Table	21
7	Words + Atomic Elements = Scientific Fact and Not Fiction	30
8	Purposely Arranged or Science Fiction? You Be the Judge	38

CHAPTER 1

The Mind of the Supreme Intelligence

"I want to know God's thoughts; the rest are details." — A. Einstein

If there is a supreme intelligence, would he think like us? Would his thought process involve collecting and analyzing information from previous experiments? Would the supreme intelligence have to perform an experiment? Is his knowledge limited, requiring periodic consultations with outside sources? What are his goals? What motivates his activities? What is he thinking right now? From our earthly position in this vast universe, it would appear that whatever his thoughts are, they would naturally be higher than ours. His understanding of physics, chemistry, biology, and the entire spectrum of our scientific knowledge would likewise be broader and deeper than our own. He would no doubt know the composition and structure of all matter, right down to the subatomic particles and below. Humans trying to comprehend the knowledge and wisdom of a supreme intelligence would be like an automotive engineer opening the hood of a car and explaining to a chimpanzee how the vehicle works. His explanation might include, "Therefore, if we extract from the ground petroleum oil and refine it through several chemical processes to an octane rating of 87 to 93, then pour it into the gas tank of the vehicle designed for that octane level, then turn the ignition key, this will create a spark in the combustion chamber of an internal combustion engine, which will turn pistons connected to the crank shaft and cause the vehicle to move forward. Chimp, does this brief explanation help?" Could the chimpanzee understand this explanation from the engineer? Of course not. The chimpanzee can be trained to put the key into the ignition and start the engine, but his mind is incapable of totally understanding how the vehicle works or how it got there. However, unlike the chimp, which will never comprehend the intricate engineering of an automobile, we can grow in our knowledge and understanding of the universe. Although we may not fully understand the mechanics of how this universe began out of nothing, we have the ability to marvel at its beauty and to acknowledge or dismiss the existence of the one who made it.

If our human brain has the ability to marvel at the world around us and process a variety of unrelated information at the same time (i.e., when driving a car) then the supreme intelligence should be able to function in a similar manner. No doubt the supreme intelligence has a thought process that is consistent with his title, a thought process that is naturally supreme in every way—multilevel, multidirectional, and multidimensional in space and time. Also, every thought may have a connecting component with every other thought. To humans, the connection among these various components may not be readily apparent and may actually appear to be totally unrelated. For example, on the human level, the connection among trees, fuel, paper, and building material is obvious; however, the connection among trees, sunlight, carbon dioxide, and oxygen is not quite that clear. Similarly, the supreme intelligence's thoughts, which are at a higher level than ours, may contain intricate connecting components that may not be clearly evident to us. To connect the individual word components of his thoughts may be like trying to "connect the dots," only at a level that is beyond the range of our normal way of thinking but yet may have some fundamental resemblance to the traditional children's game.

CHAPTER 2

Connecting the Dots

Most of us have played the game of connecting the dots, perhaps when we were kids or maybe as adults while waiting for a doctor or dental appointment. It's fun and easy to do. The objective is so obvious that game instructions are usually not provided. The dots are not randomly connected but joined from one dot to the next according to their sequential order. You start by looking for the dot next to the number one and then draw a line to the dot next to the number two, and so forth. After you have connected all the dots, the outline of a picture is revealed. Once they can identify the object, kids will usually color in the area and provide additional details omitted from the dotted outline.

This simple game of connecting the dots is designed with a starting point. All the dots are ultimately connected. Although the connections are linear, the final picture may appear to contain a curvature or some other geometric design. The principles of this game—connecting one object to the next, or one thought with the next—can be applied to other aspects of life, whether literally or figuratively. Connecting current information with previous knowledge forms the basis of all research and analysis.

Since the dawn of civilization, mankind has continuously been connecting the dots, so to speak, by connecting new things with things previously learned. We continue to apply this simple concept in virtually every aspect of our daily lives and business activities: police during criminal investigations, lawyers in building their cases, medical professionals in determining the cause of diseases—and with each endeavor we continually build on what we already know. We do it consciously and unconsciously. The impact of connecting the dots correctly may lead to better understanding, greater insight, and wisdom in handling future problems and challenges. Today, a small bit of information may appear to be insignificant; however, tomorrow that same information may be extremely valuable. The importance of connecting the dots accurately and rapidly deciphering the meaning of it all can potentially be lifesaving. Therefore, when crucial issues require immediate resolution, connecting the dots between current information and previously gathered data is not just playing a game.

The need—or failure—to connect the dots was certainly evident during the tragic events of September 11, 2001. On the surface, purchasing box cutters and attending flight-training schools have no connection. By themselves, these activities appeared to be normal, everyday events. However, today we know there was a sinister connection between those two activities.

Today, the need to connect the dots involves more than just drawing a line from one number to the next. In this extremely complex world, the answers and solutions to important questions may not be easily discernible. They might involve connecting patterns of activities and events to provide an outline of a condition or situation that we can then begin to recognize as harmless, beneficial, or potentially dangerous. This may require looking at ordinary things slightly differently. As we do, we may see something unusual within something very ordinary.

Various scientific instruments are used to help us understand the world around us and can help explain how things are connected—but these same instruments cannot tell us why things are connected. The following information will not attempt to explain *why*, but it may help readers to see that sometimes, in the most unlikely places, there is an unusual connection, one that was assuredly purposely arranged. Connecting the dots may not always occur in a straight line or with a pattern of events. There may be a new level and method of connecting the dots in a way that we thought was unrelated. Let's consider this possibility and apply this approach to another area of interest.

CHAPTER

3

Mathematical Balance

As some people already know, the numbers one through nine can be arranged so that if added horizontally, vertically, or diagonally, the results are the same: fifteen (see figure 1). This configuration of numbers dates back several thousand years and may have had some useful application among Chinese and Arab mathematicians. However, if there was some practical application to this pattern of numbers, it has long been lost. This simple arrangement, however, may contain the means for unlocking other information concerning the harmonious balance of matter that makes up the very fabric of our universe.

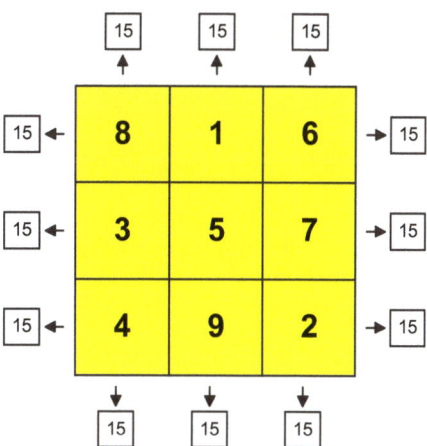

FIGURE 1

CHAPTER 4

Balancing Words with Numbers

In his quest to understand the theory of everything, Albert Einstein may have endeavored to explain theoretically and mathematically the mysteries within our universe. However, his words might reveal an underlying desire to know even more—the thoughts of the one who designed it. If the Bible is the product of the creator's thoughts, it should provide information concerning its author—his abilities and qualities.

It appears that one particular Bible verse, Galatians 5:22–23, identifies several qualities that the supreme intelligence has embodied within himself. As we will see, this verse—without being scientific and yet remaining true to science—summarizes in human terms not only his qualities but also an interesting facet of his wisdom. This verse will be the starting point for the following information and the basis for connecting the dots.

In this initial phase of connecting the dots, if we connect the nine numbers (figure 2) to the sequence of words recorded at Galatians 5:22–23 (figure 3), the results (figure 4) may mean very little to most observers. However, the end results may surprise you. The words at Galatians read in part as follows:

"Love, joy, peace, long-suffering, kindness, goodness, faith, mildness, self-control."

(Please note: Some Bible translations will render the words in these two verses differently.)

8	1	6
3	5	7
4	9	2

FIGURE 2

1 •⟷• love

2 •⟷• joy

3 •⟷• peace

4 •⟷• long-suffering

5 •⟷• kindness

6 •⟷• goodness

7 •⟷• faith

8 •⟷• mildness

9 •⟷• self-control

FIGURE 3

8 Mildness	1 Love	6 Goodness
3 Peace	5 Kindness	7 Faith
4 Long-Suffering	9 Self-Control	2 Joy

FIGURE 4

Incredible Connections

CHAPTER

5

The Analysis of Boxes 7, 4, and 1

In each game of connecting the dots, there has to be a starting point. Without a starting point, we have nothing more than various dots on a blank sheet of paper. Completing the picture accurately without additional information would virtually be impossible.

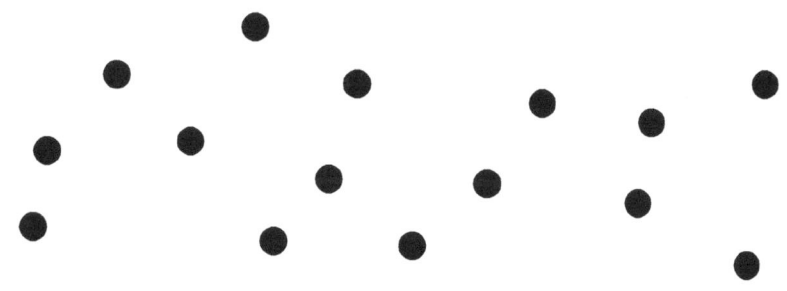

As stated previously, the starting point or the location of number one has been identified—Galatians 5:22–23. Now that we have matched a word to a number, we can now move to the next level by connecting the dots with words and not numbers. There are more than 774,000 words in the sacred writings and recorded over a period of almost 2000 years. Some words (i.e., Italian) are recorded only once. There are other words mentioned hundreds of times. For example, the word *love* is mentioned 314 times. As we will see, by connecting the right word in one verse to the same word another verse, then connecting the right word in the second verse to a similar word in the third verse, and then repeating this process, the end result will reveal scientifically that the entire sacred text had to be purposely arranged by a supreme intelligence.

Let's begin with the word *faith* in box 7 (see exhibit A). The word *faith* at Galatians 5:22 is also mentioned in connection with the word *righteous* or *righteousness* at Hebrews 10:38. The word *righteous* or *righteousness* is mentioned in connection with the word *breastplate* at Ephesians 6:14. The word *breastplate* is associated with the word *iron* at Revelations 9:9. Therefore, box 7 contains the words *faith, righteousness, breastplate,* and *iron*.

Exhibit A

8 Mildness	1 Love	6 Goodness
3 Peace	5 Kindness	7 Faith
4 Long-Suffering	9 Self-Control	2 Joy

7 — Faith, Righteousness, Breastplate, Iron

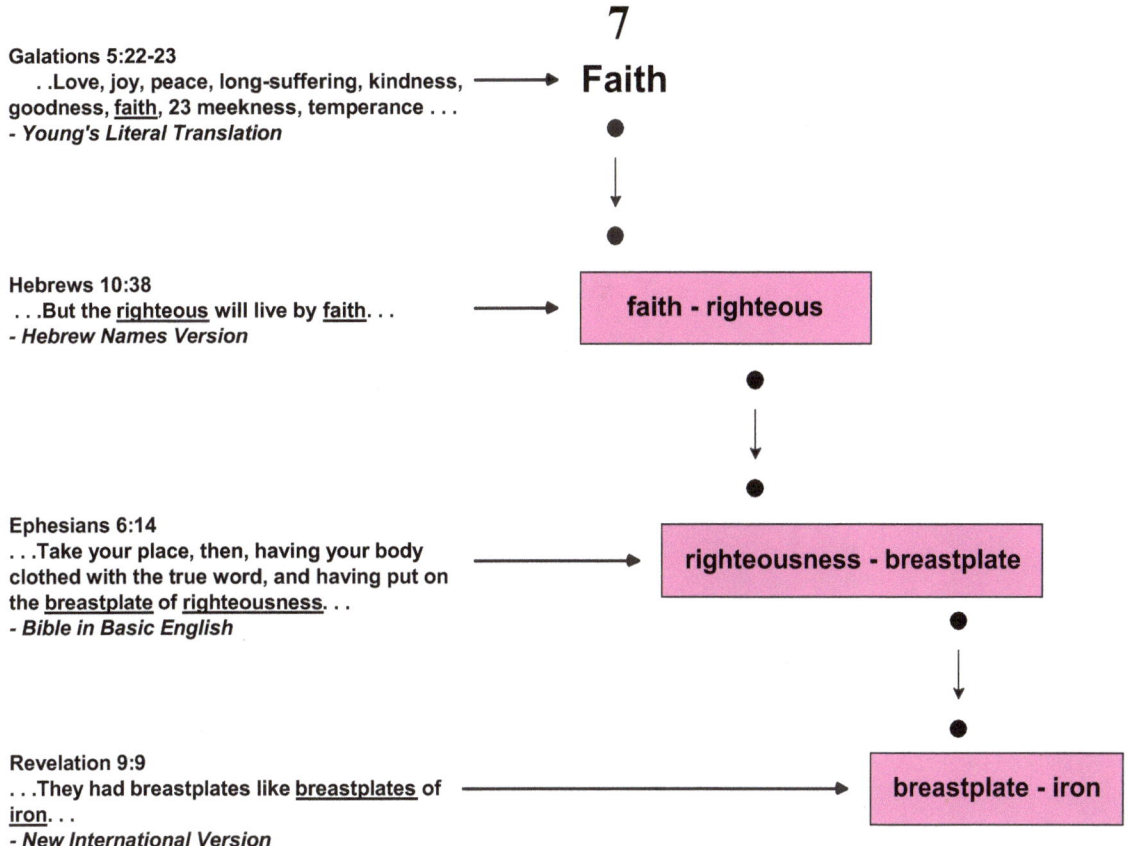

Now let's go to the word *long-suffering* recorded at Galatians 5:23 in box 4. Note that *long-suffering* is mentioned along with the word *knowledge* at 2 Corinthians 6:6; *knowledge* is connected with *gold* at Proverbs 8:10; and *gold* is connected with *silver* and *copper* at Exodus 25:3. Therefore, box 4 contains the words *long-suffering, knowledge, gold, silver*, and *copper*.

Exhibit B

8 Mildness	1 Love	6 Goodness
3 Peace	5 Kindness	7 Faith
4 Long-Suffering	9 Self-Control	2 Joy

4
Long-Suffering, Knowledge, Gold, Silver, Copper

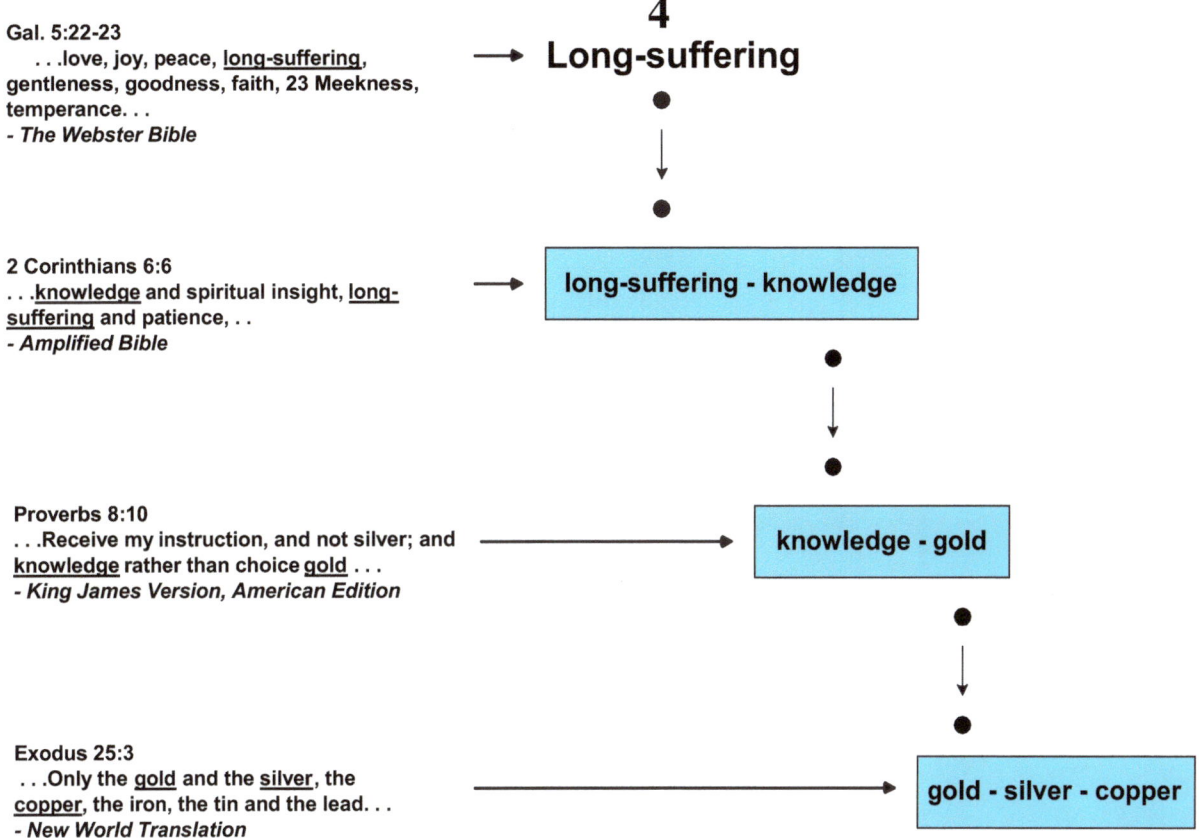

Most English Bible translations will translate the Hebrew word נחשת *nĕchosheth* or the Greek word χαλκός *chalkos* as "brass" or "bronze." However, if the following analysis is correct then the exact translation of both the Hebrew and Greek word would be "copper."

Let's do one more by examining the word *love* in box 1 as indicated in exhibit C. Again, let's start with the word *love* at Galatians 5:23. Notice that at 1 Peter 3:10, *love* is connected with the word *life*. At Genesis 2:7, the word *life* is connected with the word *breath*, and at Exodus 15:10, the word *breath* is connected with the word *lead*. At Numbers 31:22, the word *lead* is connected with the word *tin*. Therefore, box 1 contains the words *love, life, breath, lead,* and *tin*.

Exhibit C

8 Mildness	1 Love	6 Goodness
3 Peace	5 Kindness	7 Faith
4 Long-Suffering	9 Self-Control	2 Joy

1 Love, Life, Breath, Lead, Tin

1
Love

Gal. 5:22-23
...<u>love</u>, joy, peace, long-suffering, graciousness, goodness, faithfulness, meekness, self-control . . .
- *The Emphasized Bible*

→ Love

1 Peter 3:10
...For "Whoever wants to <u>love</u> <u>life</u> and see good days...
- *Complete Jewish Bible*

→ love - life

Genesis 2:7
...and breathed into his face the <u>breath</u> of <u>life</u>...
- *Douay-Rheims*

→ life - breath

Exodus 15:10
...But you blew on them with your <u>breath</u> and covered them with the sea. They sank like <u>lead</u>...
- *New Century Version*

→ breath - lead

Numbers 31:22
...Only the gold, and the silver, the copper, the iron, the <u>tin</u>, and the <u>lead</u> . . .
- *The Darby Translation*

→ lead - tin

Exhibit C
(Commentary)

Please note: At Numbers 31:22, which reads, "Only the gold and the silver, the copper, the iron, the tin and the lead," the metals gold, silver, and copper have already been identified with box 4 and iron has been linked with box 7. Concerning the remaining metals in this verse (tin and lead), there appears to be only three other occurrences in the Bible where the words *tin* and *lead* are mentioned in the same verse (see example below). It appears that *lead* is linked with *tin,* and therefore only *lead* and *tin* are placed in box 1.

Ezekiel 22:18 - "All of them are the copper, tin, iron and lead left inside a furnace."— *New International Version*

Ezekiel 22:20 - "As silver, copper, iron, lead and tin are collected in the melting-pot."— *New Jerusalem Bible*

Ezekiel 27:12 - "For silver, iron, tin, and lead, They have given out thy remnants."— *Young's Literal Translation*

CHAPTER 6

Numbers + Words = Table

On the surface it would appear that the items in boxes 7, 4, and 1 have nothing in common.

Box 7—faith, righteousness, breastplate, and iron
Box 4—long-suffering, knowledge, gold, silver, and copper
Box 1—love, life, breath, lead, and tin

The last item(s) in each box (iron, gold, silver, copper, lead, and tin) are considered metals. But let's view them not as metals but as elements—atomic elements. If we consider these metals as atomic elements and analyze this information within the framework of current scientific knowledge, the results will show that each word previously connected with that metal was not coincidentally selected and recorded by the human writer but was unmistakably and purposely arranged in that manner by the actual author.

8 Mildness	1 Love, Life, Breath, Lead, Tin	6 Goodness
3 Peace	5 Kindness	7 Faith, Righteousness, Breastplate, Iron
4 Long-Suffering, Knowledge, Gold, Silver, Copper	9 Self-Control	2 Joy

FIGURE 5

The scientific community has established a universally accepted configuration of all known atomic elements, an arrangement commonly referred to as the periodic table of elements (see figure 6). Elements are arranged based on the ranking of their characteristics.

Periodic Table of Elements

1 H Hydrogen																	2 He Helium
3 Li Lithium	4 Be Beryllium											5 B Boron	6 C Carbon	7 N Nitrogen	8 O Oxygen	9 F Fluorine	10 Ne Neon
11 Na Sodium	12 Mg Magnesium											13 Al Aluminum	14 Si Silicon	15 P Phosphorus	16 S Sulfur	17 Cl Chlorine	18 Ar Argon
19 K Potassium	20 Ca Calcium	21 Sc Scandium	22 Ti Titanium	23 V Vanadium	24 Cr Chromium	25 Mn Manganese	26 Fe Iron	27 Co Cobalt	28 Ni Nickel	29 Cu Copper	30 Zn Zinc	31 Ga Gallium	32 Ge Germanium	33 As Arsenic	34 Se Selenium	35 Br Bromine	36 Kr Krypton
37 Rb Rubidium	38 Sr Strontium	39 Y Yttrium	40 Zr Zirconium	41 Nb Niobium	42 Mo Molybdenum	43 Tc Technetium	44 Ru Ruthenium	45 Rh Rhodium	46 Pd Palladium	47 Ag Silver	48 Cd Cadmium	49 In Indium	50 Sn Tin	51 Sb Antimony	52 Te Tellurium	53 I Iodine	54 Xe Xenon
55 Cs Cesium	56 Ba Barium	57 La Lanthanum	72 Hf Hafnium	73 Ta Tantalum	74 W Tungsten	75 Re Rhenium	76 Os Osmium	77 Ir Iridium	78 Pt Platinum	79 Au Gold	80 Hg Mercury	81 Tl Thallium	82 Pb Lead	83 Bi Bismuth	84 Po Polonium	85 At Astatine	86 Rn Radon
87 Fr Francium	88 Ra Radium	89 Ac Actinium	104 Rf Rutherfordium	105 Db Dubnium	106 Sg Seaborgium	107 Bh Bohrium	108 Hs Hassium	109 Mt Meitnerium									

FIGURE 6

We can easily identify each of the metals mentioned previously—tin (Sn-50), lead (Pb-82), copper (Cu-29), silver (Ag-47), gold (Au-79), and iron (Fe-26)—on the periodic table of elements as shown in figure 7.

Periodic Table of Elements

1 H Hydrogen																	2 He Helium
3 Li Lithium	4 Be Beryllium											5 B Boron	6 C Carbon	7 N Nitrogen	8 O Oxygen	9 F Fluorine	10 Ne Neon
11 Na Sodium	12 Mg Magnesium											13 Al Aluminum	14 Si Silicon	15 P Phosphorus	16 S Sulfur	17 Cl Chlorine	18 Ar Argon
19 K Potassium	20 Ca Calcium	21 Sc Scandium	22 Ti Titanium	23 V Vanadium	24 Cr Chromium	25 Mn Manganese	26 Fe Iron	27 Co Cobalt	28 Ni Nickel	29 Cu Copper	30 Zn Zinc	31 Ga Gallium	32 Ge Germanium	33 As Arsenic	34 Se Selenium	35 Br Bromine	36 Kr Krypton
37 Rb Rubidium	38 Sr Strontium	39 Y Yttrium	40 Zr Zirconium	41 Nb Niobium	42 Mo Molybdenum	43 Tc Technetium	44 Ru Ruthenium	45 Rh Rhodium	46 Pd Palladium	47 Ag Silver	48 Cd Cadmium	49 In Indium	50 Sn Tin	51 Sb Antimony	52 Te Tellurium	53 I Iodine	54 Xe Xenon
55 Cs Cesium	56 Ba Barium	57 La Lanthanum	72 Hf Hafnium	73 Ta Tantalum	74 W Tungsten	75 Re Rhenium	76 Os Osmium	77 Ir Iridium	78 Pt Platinum	79 Au Gold	80 Hg Mercury	81 Tl Thallium	82 Pb Lead	83 Bi Bismuth	84 Po Polonium	85 At Astatine	86 Rn Radon
87 Fr Francium	88 Ra Radium	89 Ac Actinium	104 Rf Rutherfordium	105 Db Dubnium	106 Sg Seaborgium	107 Bh Bohrium	108 Hs Hassium	109 Mt Meitnerium									

FIGURE 7

Now, let's indicate the box number that is associated with each element (see figure 8).

Periodic Table of Elements

1 H Hydrogen																	2 He Helium
3 Li Lithium	4 Be Beryllium											5 B Boron	6 C Carbon	7 N Nitrogen	8 O Oxygen	9 F Fluorine	10 Ne Neon
11 Na Sodium	12 Mg Magnesium											13 Al Aluminum	14 Si Silicon	15 P Phosphorus	16 S Sulfur	17 Cl Chlorine	18 Ar Argon
19 K Potassium	20 Ca Calcium	21 Sc Scandium	22 Ti Titanium	23 V Vanadium	24 Cr Chromium	25 Mn Manganese	26 Fe Iron	27 Co Cobalt	28 Ni Nickel	29 Cu Copper	30 Zn Zinc	31 Ga Gallium	32 Ge Germanium	33 As Arsenic	34 Se Selenium	35 Br Bromine	36 Kr Krypton
37 Rb Rubidium	38 Sr Strontium	39 Y Yttrium	40 Zr Zirconium	41 Nb Niobium	42 Mo Molybdenum	43 Tc Technetium	44 Ru Ruthenium	45 Rh Rhodium	46 Pd Palladium	47 Ag Silver	48 Cd Cadmium	49 In Indium	50 Sn Tin	51 Sb Antimony	52 Te Tellurium	53 I Iodine	54 Xe Xenon
55 Cs Cesium	56 Ba Barium	57 La Lanthanum	72 Hf Hafnium	73 Ta Tantalum	74 W Tungsten	75 Re Rhenium	76 Os Osmium	77 Ir Iridium	78 Pt Platinum	79 Au Gold	80 Hg Mercury	81 Tl Thallium	82 Pb Lead	83 Bi Bismuth	84 Po Polonium	85 At Astatine	86 Rn Radon
87 Fr Francium	88 Ra Radium	89 Ac Actinium	104 Rf Rutherfordium	105 Db Dubnium	106 Sg Seaborgium	107 Bh Bohrium	108 Hs Hassium	109 Mt Meitnerium									

FIGURE 8

7 4 1

Now that we have identified on the periodic table the elements related to boxes 7, 4, and 1, it may be safe to assume that boxes 2, 3, 5, 6, 8, and 9 in figure 9 may also contain elements with corresponding atomic values.

8 Mildness	1 Love Breath of life Lead, Tin	6 Goodness
3 Peace	5 Kindness	7 Faith Breastplate Righteousness Iron
4 Long-Suffering Knowledge Gold, Silver Copper	9 Self-Control	2 Joy

FIGURE 9

In figure 7, we connected the six elements to the periodic table. Now, we have just enough information to achieve something extremely interesting. As you may remember, in the game of connecting the dots you cannot draw a line from dot number one to number four then to number seven. The rules of the game state that lines must be drawn in sequential order. In this case, how do we identify and locate the missing numbers? A casual view of the circled elements in figure 8 and the corresponding box numbers, reading left to right (7, 4, and 1), provides little information. However, if we read the box numbers in *reverse*, or right to left, we are able to fill in the missing gaps (boxes 2, 3, 5, 6, 8, and 9) as indicated in figure 10.

Periodic Table of Elements

FIGURE 10

An assumption was made that if box 4 has three elements then all boxes (1–9) would contain three elements. If our assumptions are correct, and we tally the atomic number of each element, then the horizontal and vertical totals should agree as in figure 11.

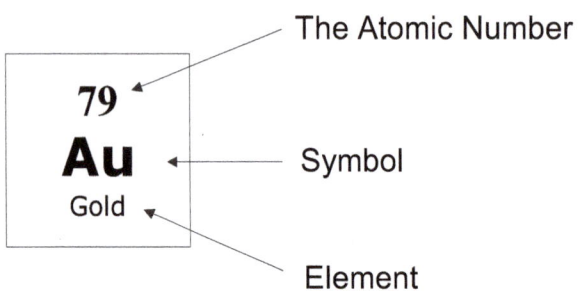

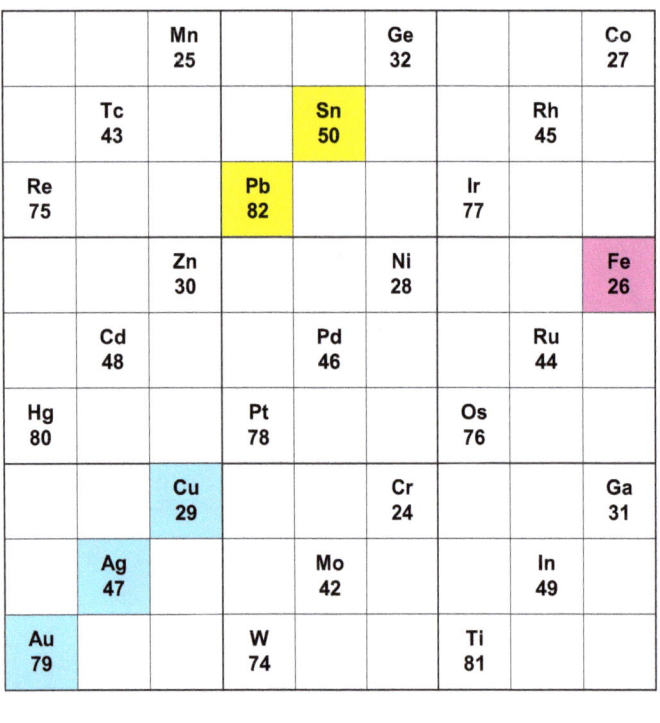
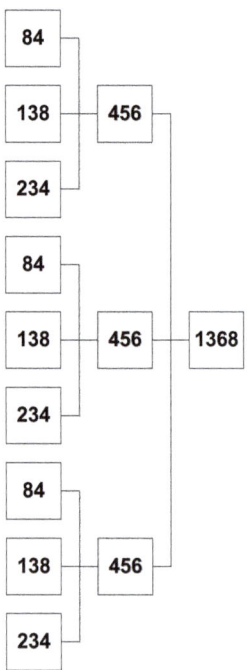
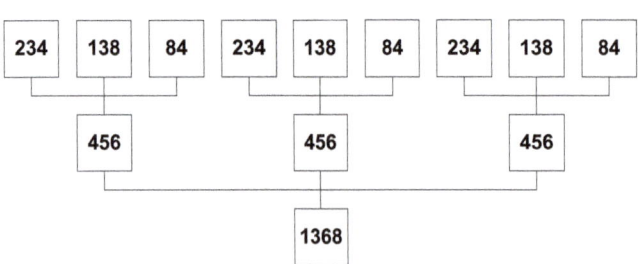

FIGURE 11

We are now left with a periodic table that looks like figure 12 below.

Periodic Table of Elements

1 H Hydrogen																	2 He Helium
3 Li Lithium	4 Be Beryllium											5 B Boron	6 C Carbon	7 N Nitrogen	8 O Oxygen	9 F Fluorine	10 Ne Neon
11 Na Sodium	12 Mg Magnesium											13 Al Aluminum	14 Si Silicon	15 P Phosphorus	16 S Sulfur	17 Cl Chlorine	18 Ar Argon
19 K Potassium	20 Ca Calcium	21 Sc Scandium	22 Ti Titanium	23 V Vanadium								33 As Arsenic	34 Se Selenium	35 Br Bromine	36 Kr Krypton		
37 Rb Rubidium	38 Sr Strontium	39 Y Yttrium	40 Zr Zirconium	41 Nb Niobium								51 Sb Antimony	52 Te Tellurium	53 I Iodine	54 Xe Xenon		
55 Cs Cesium	56 Ba Barium	57 La Lanthanum	72 Hf Hafnium	73 Ta Tantalum								83 Bi Bismuth	84 Po Polonium	85 At Astatine	86 Rn Radon		
87 Fr Francium	88 Ra Radium	89 Ac Actinium	104 Rf Rutherfordium	105 Db Dubnium	106 Sg Seaborgium	107 Bh Bohrium	108 Hs Hassium	109 Mt Meitnerium									

FIGURE 12

If we take the remaining atomic elements in figure 12 and assume that a few elements are missing (negative three, negative two, negative one, and zero) and include them in figure 11 and follow the same sequential pattern, then the total of all elements (by adding their atomic numbers) should mathematically agree both horizontally and vertically (see figure 13).

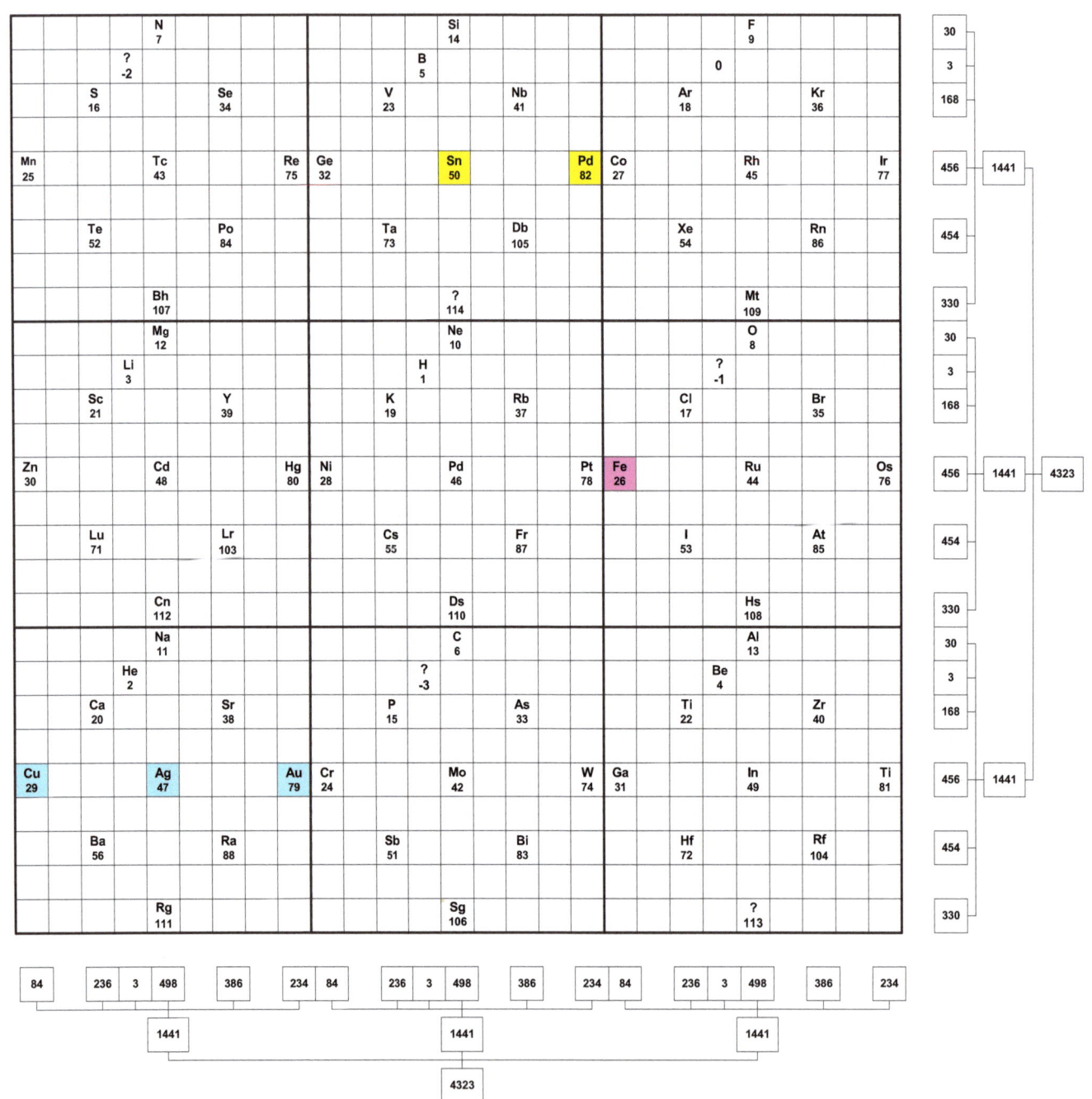

FIGURE 13

At this point, we can observe the order and balance of all known elements at the atomic level within one unified configuration. It's possible this arrangement and the position each element has in relationship to other elements may hold the key to unlocking questions puzzling scientists today.

This unique analysis of connecting the dots started with a simple concept applied in an unconventional manner. Further steps along this line of thought may be far more complicated than comprehending the connection between petroleum oil and the internal combustion engine.

CHAPTER

7

Words + Atomic Elements = Scientific Fact and Not Fiction

Let's review for a moment. I believe at this point we can assume that *every* word in the sacred text is precisely chosen and not just randomly selected to complete a thought. Like the many nerve cells (neurons) in the brain, each word has an intricate link with other words, whether previously recorded or mentioned later. So far, the word connection establishes a link with science, chemistry, and atomic elements, but does this analysis agree with scientific fact?

We will now analyze this information further by connecting one of the dots in box 7—but this time in another direction. Again, let's start with the word *faith* at Galatians 5:23. Please note that at Romans 3:25, *faith* is mentioned in connection with the word *blood*. Therefore, in box 7 we have the words *faith* and *blood*. Now let's match the atomic elements connected with box 7. Most medical professionals agree that iron (Fe-26) and oxygen (O-8) are essential elements of human blood.

Exhibit D

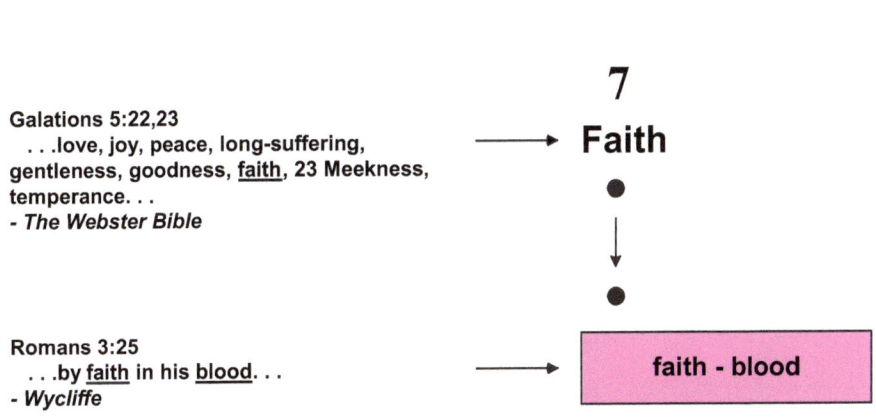

Galations 5:22,23
...love, joy, peace, long-suffering, gentleness, goodness, faith, 23 Meekness, temperance...
- *The Webster Bible*

Romans 3:25
...by faith in his blood...
- *Wycliffe*

FIGURE 14

As we look more closely at box 4, we will continue connecting the dots following the last items (gold, silver, and copper) in exhibit B on page 16. As indicated below, box 4 now includes flesh, bones, and teeth plus those items mentioned earlier in exhibit B. It's of interest to note that in figure 15 the elements sodium (Na-11), calcium (Ca-20), and strontium (Sr-38) are also included in box 4, and these items are essential elements for the development of bones and teeth.

Exhibit B (continued)

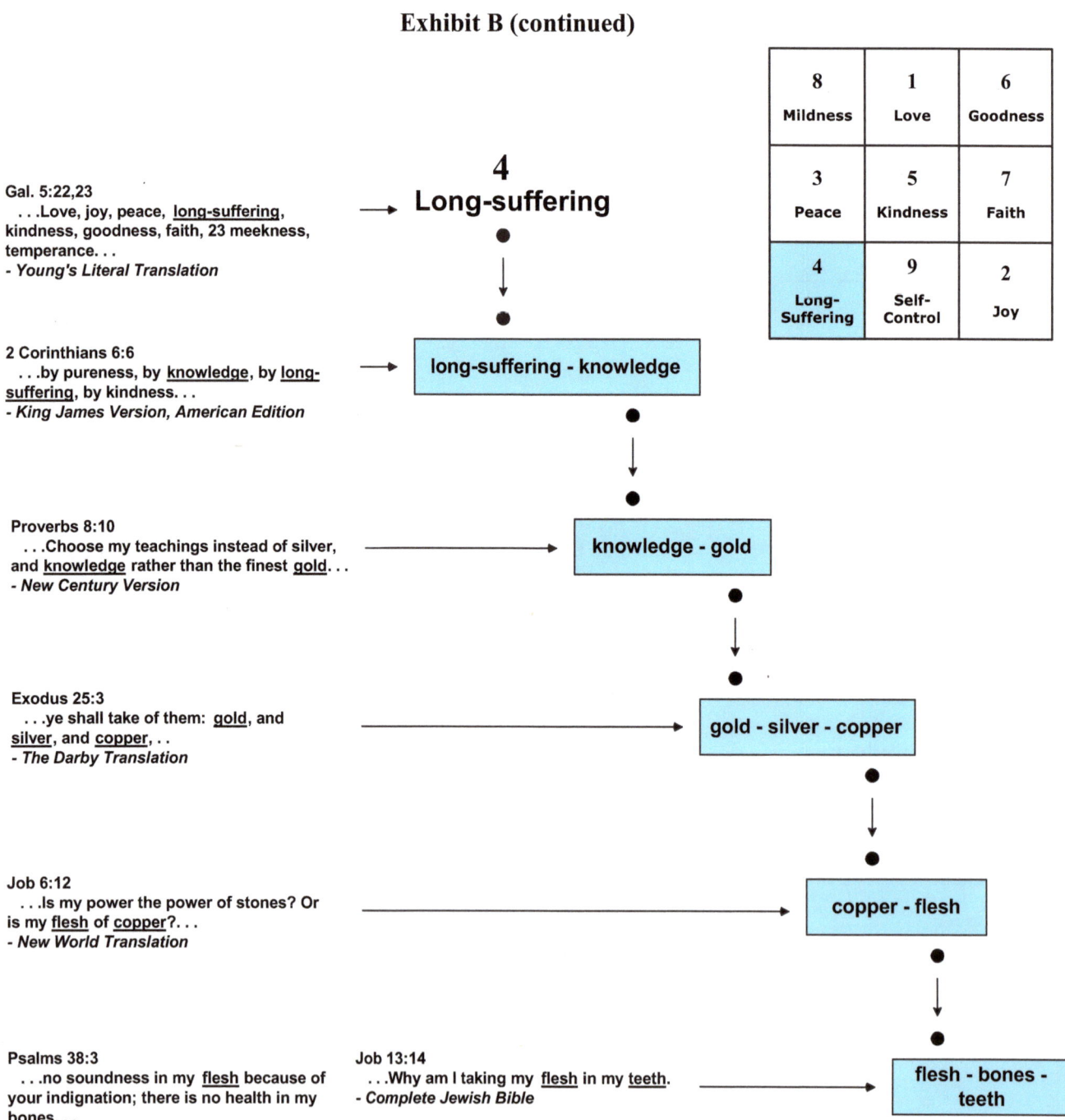

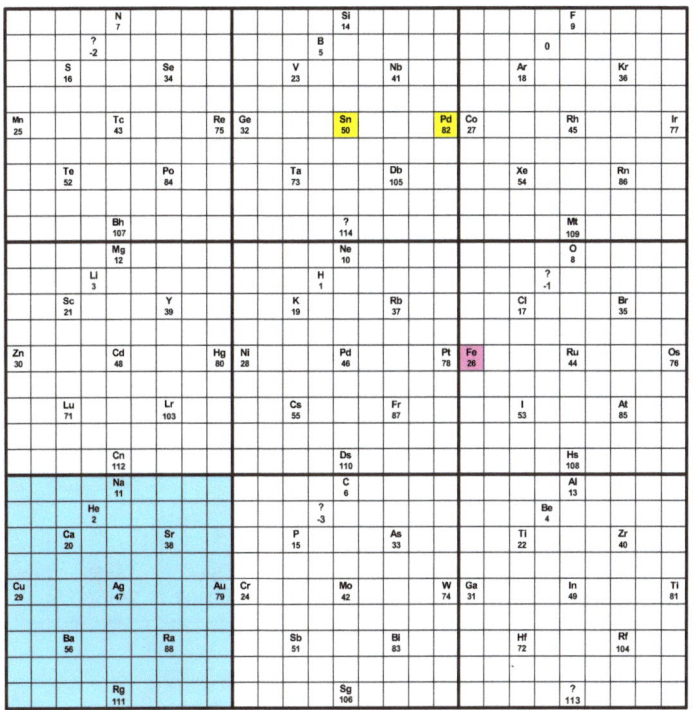
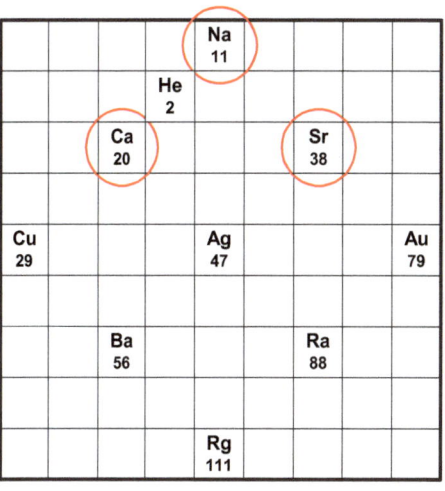

FIGURE 15

The next analysis looks at box 5. Again, consider the word *kindness* mentioned at Galatians 5:22. Notice that the word *kindness* or *loving-kindness* and the words *olive tree* are mentioned at Psalms 52:8. The words *kindness* and *olive tree* are now connected with box 5, which also contains four alkali elements (see figure 16). Many olive trees do very well in alkali or alkaline soil. This type of soil gets its name from the alkali group of atomic elements. Note that four of the six alkali elements—potassium (K-19), rubidium (Rb-37), cesium (Cs-55), and francium (Fr-87)—are included in box 5.

Exhibit E

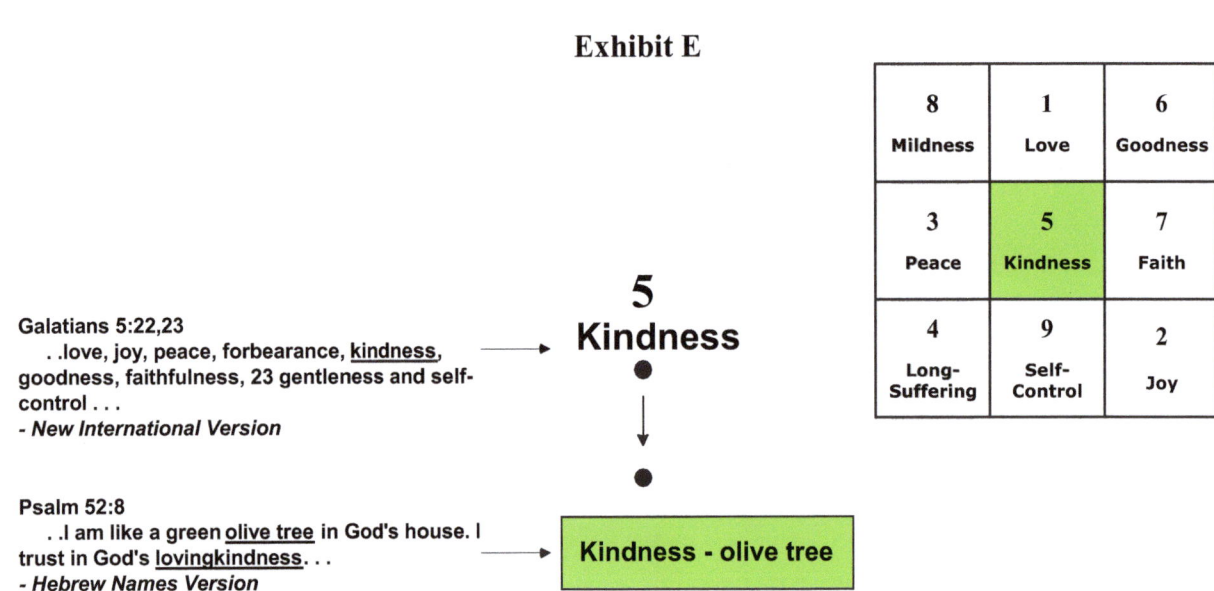

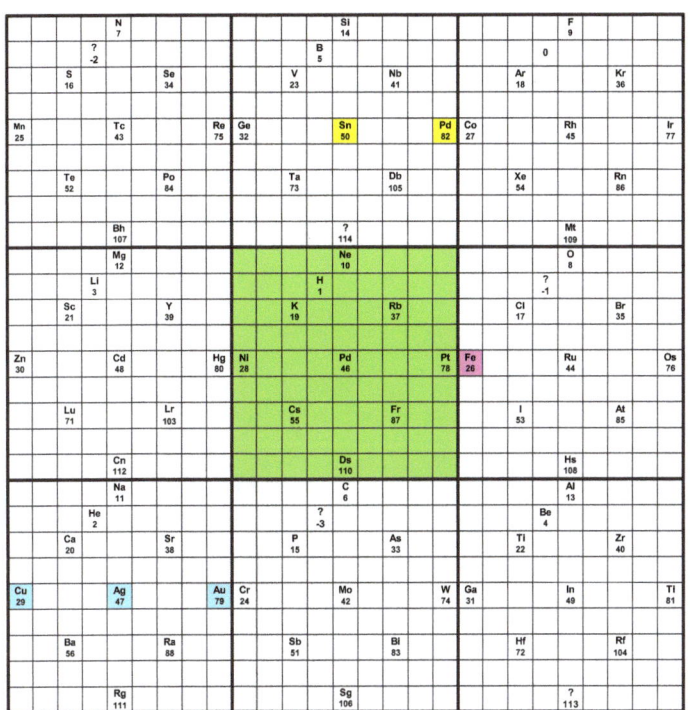
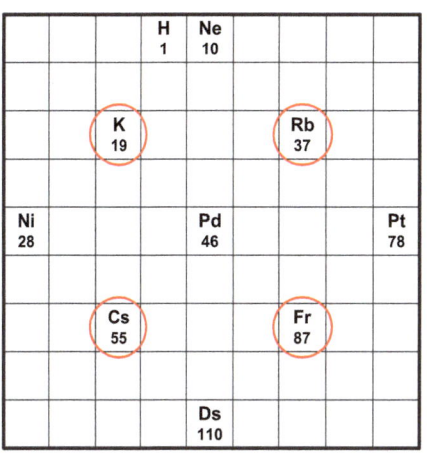

FIGURE 16

The final analysis will look at boxes 4 and 8. As shown below, the last items listed in box 4 are milk, honey, and wheat. Box 8 also ends with the item honey. Perhaps only a nutritionist would appreciate the significance of boxes 4 and 8. It's of interest to note that each of these three food items may represent at the nutrient level fat (milk), protein (honey), and carbohydrates (wheat). Some of the atomic elements associated with boxes 4 and 8 appear to support this conclusion.

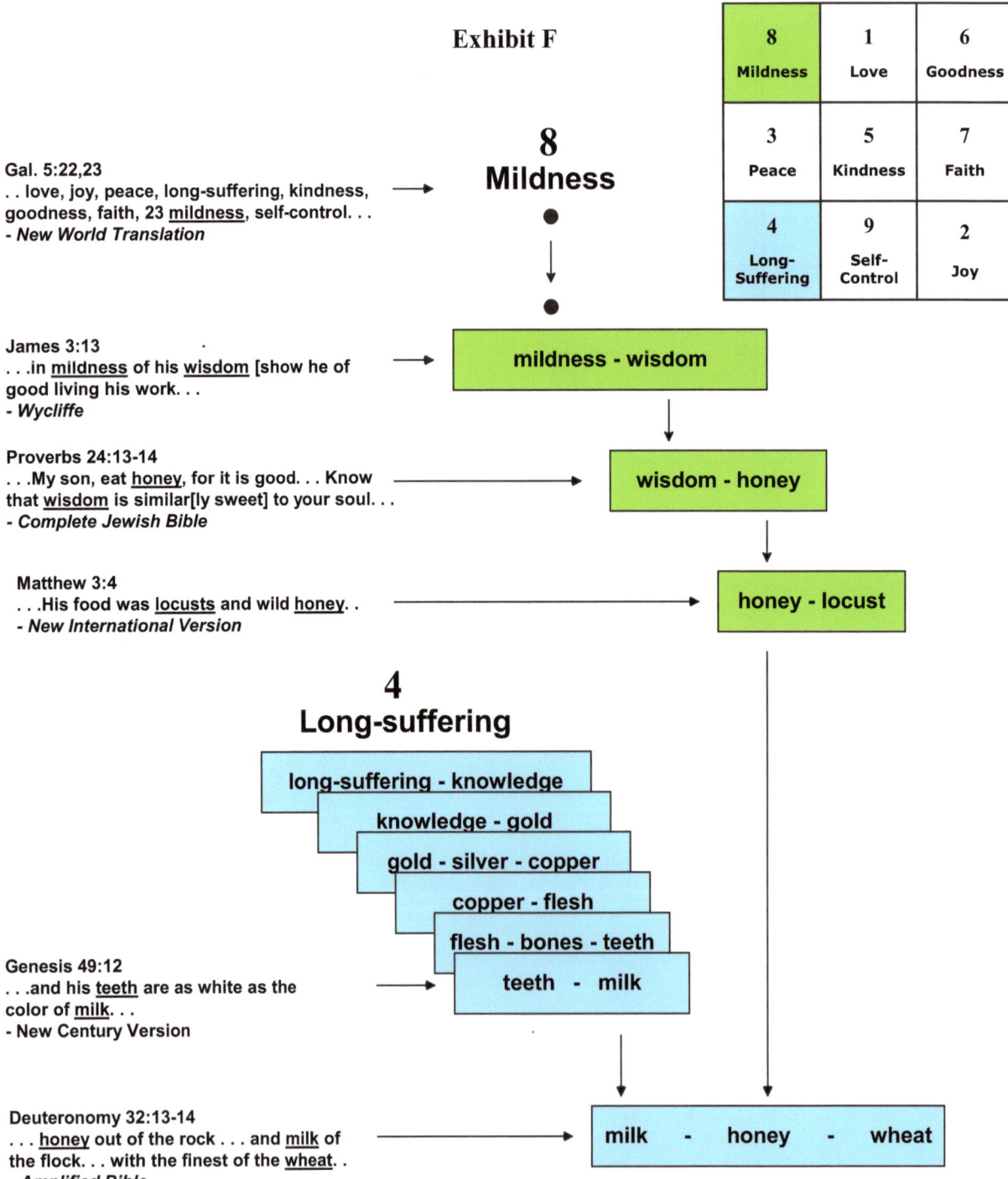

FIGURE 17

Box 8

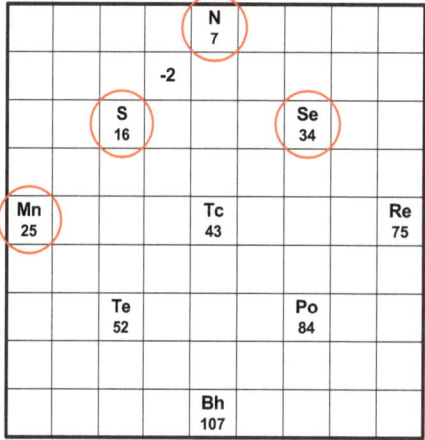

Box 4

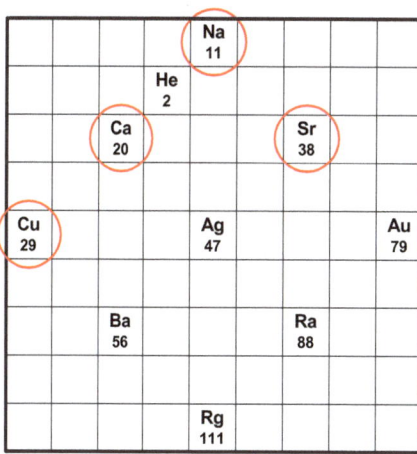

CHAPTER

8

Purposely Arranged or Science Fiction? You Be the Judge

Is it possible to replicate this analysis with any other book in the world—past or present—and achieve the same results? What are the odds of someone finding this information? This brief analysis supports the notion that in every instance, the supreme intelligence did not select words at random in order to record the greatest story ever told. Rather, he used the precise word in each sentence and knew the exact connection each word would have with other words that were later used. Although connecting words in this manner is not necessary to the overall spiritual message, nevertheless each word selected has made it possible (using figure 1) to figuratively connect the dots. Through this unusual method of connecting the dots, we have seen something that was never visible before and, in the process, have come to appreciate the unique wisdom of the supreme intelligence.

This analysis may do very little to fortify the faith of those individuals who already believe in a supreme being. For you, faith in the creator and his inspired word is not based on some interesting analysis of dots, circles, and squares. People who believe in a supreme intelligence are convinced of his existence without such information. This analysis only serves to confirm what you already believe.

But to those who claim that belief in a divine being is not provable and incompatible with current scientific thinking, may this analysis provide convincing proof not only of his existence but his incredible wisdom. May you approach the entire sacred text not with skepticism but with profound awe and hopefully will acknowledge that every word was purposely selected by the author. As the next book will demonstrate, there is more amazing information contained within those sacred words than you can ever imagine. Truly, only the supreme intelligence could skillfully weave pieces of scientific fact—at the atomic level—within the greatest story ever told.

List of all twenty-four verses previously quoted along with additional translations of that verse.

No. 1
Genesis 2:7
(See exhibit C.)

"Breathing into him the breath of life."
— *Bible in Basic English*

"And breathed into his nostrils the breath of life."
— *Complete Jewish Bible*

"And breathed into his face the breath of life."
— *Douay-Rheims*

"He breathed life-giving breath into his nostrils."
— *Good News Bible*

"And breathed into his nostrils the breath of life."
— *Hebrew Names Version*

"He breathed the breath of life."
— *New Century Version*

"And breathed into his nostrils the breath of life."
— *New International Version*

"Breathed into his nostrils the breath of life."
— *Revised Standard Version*

"And breathed into his nostrils the breath of life."
— *The Darby Translation*

"And breathed in his nostrils the breath of life."
— *The Emphasized Bible*

"And breathed into his nostrils the breath of life."
— *The Webster Bible*

"And breathed into his face the breathing of life."
— *Wycliffe*

No. 2
Genesis 49:12
(See exhibit F.)

"His eyes shall be red with wine, And his teeth white with milk."
— *American Standard Version*

"His teeth whiter than milk."
— *Complete Jewish Bible*

"Your teeth whiter than milk."
— *Contemporary English Version*

"And his teeth whiter than milk."
— *Douay-Rheims*

"His teeth white with milk."
— *Hebrew Names Version*

"And his teeth are as white as the color of milk."
— *New Century Version*

"Whiteness of his teeth is from milk."
— *New World Translation*

"His eyes shall be red with wine, and his teeth white with milk."
— *Revised Standard Version*

"And the teeth [are] white with milk."
— *The Darby Translation*

"Whiter—his teeth than milk!"
— *The Emphasized Bible*

"And his teeth be whiter than milk."
— *Wycliffe*

"And white [are] teeth with milk!"
— *Young's Literal Translation*

No. 3
Exodus 15:10
(See exhibit C.)

"Thou didst blow with thy wind, the sea covered them: They sank as lead in the mighty waters."
— *American Standard Version*

"You blew with your wind; the sea covered over them. They sank like lead in the towering waters."
— *Common English Bible*

"You blew with your wind, the sea covered them, they sank like lead in the mighty waters."
— *Complete Jewish Bible*

"Your breath blew the sea over them. They sank like lead."
— *God's Word Translation*

"But one breath from you, LORD, and the Egyptians were drowned; they sank like lead."
— *Good News Bible*

"But You blew with Your breath, and the sea covered them. They sank like lead."
— *Holman Christian Standard*

"But you blew on them with your breath and covered them with the sea. They sank like lead."
— *New Century Version*

"You blew with your breath, and the sea covered them. They sank like lead."
— *New International Version*

"You blew with your breath, the sea closed over them; they sank like lead."
— *New Jerusalem Bible*

"But with a blast of your breath, the sea covered them. They sank like lead."
— *New Living Translation*

"Thou didst blow with thy wind, the sea covered them: they sunk as lead in the mighty waters."
— *The Webster Bible*

"Spirit blew (Thou blewest with thy breath), and the sea covered them; they were drowned as lead."
— *Wycliffe*

No. 4
Exodus 25:3
(See exhibits B; B (continued))

"This is the offering you shall receive from them: gold, silver, and bronze."
— *Amplified Bible*

"Here is a list of what you are to collect: Gold, silver, and bronze."
— *Contemporary English Version*

"And this is the contribution that you shall receive from them: gold, silver, and bronze."
— *English Standard Version*

"This is the kind of contribution you will accept from them: gold, silver, and bronze."
— *God's Word Translation*

"These offerings are to be: gold, silver, and bronze."
— *Good News Bible*

"And this is the offering which ye shall take of them; gold, and silver, and brass."
— *King James Version*

"This is the contribution which you are to raise from them: gold, silver and bronze."
— *New American Standard Bible*

"These are the offerings you are to receive from them: gold, silver and bronze."
— *New International Version*

"And this is what you will accept from them: gold, silver and bronze."
— *New Jerusalem Bible*

"And this is the contribution that YOU are to take up from them: gold and silver and copper."
— *New World Translation*

"Ye shall take of them: gold, and silver, and copper."
— *The Darby Translation*

"Ye shall take of them, — gold and silver and bronze."
— *The Emphasized Bible*

No. 5
Numbers 31:22
(See exhibit C.)

"Howbeit the gold, and the silver, the brass, the iron, the tin, and the lead."
— *American Standard Version*

"But gold and silver and brass and iron and tin and lead."
— *Bible in Basic English*

"Gold, silver, copper, iron, tin, and lead."
— *Common English Bible*

"Even though gold, silver, brass, iron, tin and lead."
— *Complete Jewish Bible*

"Gold, and silver, and brass, and iron, and lead, and tin."
— *Douay-Rheims*

"Only the gold, the silver, the bronze, the iron, the tin, and the lead."
— *English Standard Version*

"Any gold, silver, bronze, iron, tin, or lead."
— *God's Word Translation*

"Only the gold, and the silver, the brass, the iron, the tin, and the lead."
— *King James Version*

"Only the gold and the silver, the copper, the iron, the tin and the lead."
— *New World Translation*

"Only the gold, the silver, the bronze, the iron, the tin, and the lead."
— *Revised Standard Version*

"Only the gold, and the silver, the copper, the iron, the tin, and the lead."
— *The Darby Translation*

"Only, the gold, and the silver, the brass, the iron, the tin, and the lead."
— *Young's Literal Translation*

No. 6
Deuteronomy 32:13–14
(See exhibit F.)

"Honey out of the rock ... and milk of the flock ... with the finest of the wheat."
— *Amplified Bible*

"Suck honey from the rocks ... milk from the sheep ... with the finest wheat flour."
— *Complete Jewish Bible*

"And honey was found among the rocks.... And herds produced milk.... Your wheat was the finest."
— *Contemporary English Version*

"Honey out of the rock ... and milk from the flock ... with the very finest of the wheat."
— *English Standard Version*

"Honey among the rocks ... cows and goats gave plenty of milk ... and cattle, the finest wheat."
— *Good News Bible*

"Honey from the rock and ... herd and milk from the flock ... with the choicest grains of wheat."
— *Holman Christian Standard*

"Honey from the rock ... and milk of the flock ... With the finest of the wheat."
— *New American Standard Bible*

"Honey from the rocks ... milk from the flock ... and the best of the wheat."
— *New Century Version*

"Honey from the rock ... milk from herd and flock ... and the finest kernels of wheat."
— *New International Version*

"Honey out of the cliff ... milk of sheep ... kernels of wheat."
— *The Emphasized Bible*

"Honey out of the rock ... and milk of sheep ... with the fat of kidneys of wheat."
— *The Webster Bible*

"Honey out of the rock ... and milk of the flock ... With the finest of the wheat."
— *World English Bible*

No. 7
Job 6:12
(See exhibit B (continued))

"Is my strength and endurance that of stones? Or is my flesh made of bronze?"
— ***Amplified Bible***

"Is my strength the strength of stones, or is my flesh brass?"
— ***Bible in Basic English***

"Is my strength the strength of stones? Is my flesh made of bronze?"
— ***Complete Jewish Bible***

"Am I made of stone? Is my body bronze?"
— ***Good News Bible***

"Is my strength the strength of stones? Or is my flesh of brass?"
— ***King James Version, American Edition***

"Is my strength the strength of stones, Or is my flesh bronze?"
— ***New American Standard Bible***

"Is my flesh bronze?"
— ***New International Version***

"Is mine the strength of stone, is my flesh made of bronze?"
— ***New Jerusalem Bible***

"Do I have strength as hard as stone? Is my body made of bronze?"
— ***New Living Translation***

"Is my power the power of stones? Or is my flesh of copper?"
— ***New World Translation***

"Is my strength the strength of stones? Is my flesh of brass?"
— ***The Darby Translation***

"Or is, my flesh, of bronze?"
— ***The Emphasized Bible***

No. 8
Job 13:14
(See exhibit B (continued))

"Wherefore should I take my flesh in my teeth, And put my life in my hand?"
— *American Standard Version*

"Why should I take my flesh in my teeth and put my life in my hands?"
— *Amplified Bible*

"Why am I taking my flesh in my teeth?"
— *Complete Jewish Bible*

"Why do I tear my flesh with my teeth?"
— *Douay-Rheims*

"Why should I take my flesh in my teeth and put my life in my hand?"
— *English Standard Version*

"Why do I put myself at risk and take my life in my own hands?"
— *Holman Christian Standard*

"Wherefore do I take my flesh in my teeth, and put my life in mine hand?"
— *King James Version, American Edition*

"Why should I take my flesh in my teeth And put my life in my hands?"
— *New American Standard Bible*

"Should I take my flesh in my teeth, and put my life in my hand?"
— *The Darby Translation*

"I will take up my flesh in my teeth."
— *The Emphasized Bible*

"Why rend I my flesh with my teeth."
— *Wycliffe*

"Wherefore do I take my flesh in my teeth?"
— *Young's Literal Translation*

No. 9
Psalms 38:3
(See exhibit B (continued))

"No soundness in my flesh…. Your indignation; neither is there any health or rest in my bones."
— *Amplified Bible*

"There's nothing in my body that isn't broken because of your rage; there's no health in my bones."
— *Common English Bible*

"There is no health in my flesh, because of thy wrath: there is no peace for my bones."
— *Douay-Rheims*

"No soundness in my flesh because of your indignation; there is no health in my bones."
— *English Standard Version*

"No soundness in my flesh because of your indignation, Neither is there any health in my bones."
— *Hebrew Names Version*

"No soundness in my flesh because of thine anger; neither is there any rest in my bones."
— *King James Version, American Edition*

"No soundness in my flesh because of Your indignation; There is no health in my bones."
— *New American Standard Bible*

"No sound spot in my flesh because of your denunciation. There is no peace in my bones."
— *New World Translation*

"There is no soundness in my flesh because of thy indignation; there is no health in my bones."
— *Revised Standard Version*

"No soundness in my flesh because of thine indignation; no peace in my bones."
— *The Darby Translation*

"None health is in my flesh from the face of thine ire; no peace is to my bones."
— *Wycliffe*

"Soundness is not in my flesh, Because of Thine indignation, Peace is not in my bones."
— *Young's Literal Translation*

No. 10
Psalm 52:8
(See exhibit E.)

"I am like a green olive tree in the house of God; I trust in and confidently rely on the lovingkindness and the mercy of God forever and ever."
— *Amplified Bible*

"I am like a branching olive-tree in the house of God; I have put my faith in his mercy for ever and ever."
— *Bible in Basic English*

"I am like a leafy olive tree in the house of God; I put my trust in the grace of God forever and ever."
— *Complete Jewish Bible*

"I, as a fruitful olive tree in the house of God, have hoped in the mercy of God for ever, yea for ever and ever."
— *Douay-Rheims*

"I am like a green olive tree in God's house. I trust in God's lovingkindness."
— *Hebrew Names Version*

"I am like a green olive tree in the house of God; I trust in the lovingkindness of God forever and ever."
— *New American Standard Bible*

"Like a flourishing olive tree in the house of God, put my trust in God's faithful love, for ever and ever."
— *New Jerusalem Bible*

"But I shall be like a luxuriant olive tree in God's house; I do trust in the loving-kindness of God to time indefinite, even forever."
— *New World Translation*

"Flourishing olive-tree, in the house of God, I have put confidence in the lovingkindness of God."
— *The Emphasized Bible*

"But I [am] like a green olive tree in the house of God: I trust in the mercy of God for ever and ever."
— *The Webster Bible*

"I am like a green olive tree in God's house. I trust in God's lovingkindness forever and ever."
— *World English Bible*

"As a green olive in the house of God, I have trusted in the kindness of God."
— *Young's Literal Translation*

No. 11
Proverbs 8:10
(See exhibits B; B (continued))

"Receive my instruction, and not silver; And knowledge rather than choice gold."
— *American Standard Version*

"My instruction in preference to [striving for] silver, and knowledge rather than choice gold."
— *Amplified Bible*

"Receive my instruction, rather than silver; knowledge, rather than the finest gold."
— *Complete Jewish Bible*

"Take my instruction instead of silver, and knowledge rather than choice gold."
— *English Standard Version*

"Choose my instruction instead of silver; choose knowledge rather than the finest gold."
— *Good News Bible*

"Receive my instruction, and not silver; and knowledge rather than choice gold."
— *King James Version, American Edition*

"Take my instruction and not silver, And knowledge rather than choicest gold."
— *New American Standard Bible*

"Choose my teachings instead of silver, and knowledge rather than the finest gold."
— *New Century Version*

"Choose my instruction instead of silver, knowledge rather than choice gold."
— *New International Version*

"Accept my discipline rather than silver, and knowledge of me in preference to finest gold."
— *New Jerusalem Bible*

"Receive my instruction, and not silver; and knowledge rather than choice gold."
— *The Darby Translation*

"Receive my instruction, and not silver, And knowledge rather than choice gold."
— *Young's Literal Translation*

No. 12
Proverbs 24:13–14
(See exhibit F.)

"My son, eat honey, for it is good … Know that wisdom is similar[ly sweet] to your soul."
— ***Complete Jewish Bible***

"Eat honey, my son, because it is good … So also is the doctrine of wisdom to thy soul."
— ***Douay-Rheims***

"Eat honey, for it is good, and … Know that wisdom is such to your soul."
— ***English Standard Version***

"Honey from the comb is sweet on your tongue, you may be sure that wisdom is good for the soul."
— ***Good News Bible***

"Eat thou honey, because it is good … so shall the knowledge of wisdom be unto thy soul."
— ***King James Version, American Edition***

"Honey from the comb is sweet to your taste. Know also that wisdom is like honey for you."
— ***New International Version***

"Eat honey, my child, since it is good … and so, for sure, will wisdom be to your soul."
— ***New Jerusalem Bible***

"Let sweet comb honey be upon your palate. In the same way, do know wisdom for your soul."
— ***New World Translation***

"Honey, my son, for it is good;… honeycomb is sweet to thy taste: so consider wisdom for thy soul."
— ***The Darby Translation***

"Eat thou honey, because it is good.… Thus, take knowledge of wisdom."
— ***The Emphasized Bible***

"Eat thou honey, for it is good;… So and the teaching of wisdom is good to thy soul."
— ***Wycliffe***

"Honey that [is] good,… So [is] the knowledge of wisdom to thy soul."
— ***Young's Literal Translation***

No. 13
Ezekiel 22:18
(See exhibit C (commentary))

"All of them are brass and tin and iron and lead, in the midst of the furnace."
— *American Standard Version*

"All of them are bronze and tin and iron and lead in the midst of the furnace."
— *Amplified Bible*

"All copper, tin, iron, and lead. In the furnace, they've become the waste product of silver."
— *Common English Bible*

"As worthless as the leftover metal in a furnace after silver has been purified."
— *Contemporary English Version*

"All of them are like copper, tin, iron, and lead in a smelting furnace."
— *God's Word Translation*

"All of them are copper, tin, iron, and lead inside the furnace."
— *Holman Christian Standard Bible*

"All they [are] brass, and tin, and iron, and lead, in the midst of the furnace."
— *King James Version*

"They are like the copper, tin, iron, and lead left in the furnace."
— *New Century Version*

"All of them, silver and bronze and tin and iron and lead in the furnace."
— *Revised Standard Version*

"They are all copper, and tin, and iron, and lead, in the midst of the furnace."
— *The Darby Translation*

"They [are] brass, and tin, and iron, and lead, in the midst of the furnace."
— *The Webster Bible*

"All of them are brass and tin and iron and lead."
— *World English Bible*

No. 14
Ezekiel 22:20
(See exhibit C (commentary))

"Just as silver, copper, iron, lead, and tin are collected and placed in a furnace to fan the flames."
— ***Common English Bible***

"Then, just as they collect silver, copper, iron, lead and tin into a crucible and blow fire on it."
— ***Complete Jewish Bible***

"In the same way that the ore of silver, copper, iron, lead, and tin is put in a refining furnace."
— ***Good News Bible***

"As they gather silver and brass and iron and lead and tin into the midst of the furnace."
— ***Hebrew Names Version***

"Just as one gathers silver, copper, iron, lead, and tin into the furnace."
— ***Holman Christian Standard***

"People put silver, copper, iron, lead, and tin together inside a furnace."
— ***New Century Version***

"As silver, copper, iron, lead and tin are gathered into a furnace to be melted."
— ***New International Version***

"As silver, copper, iron, lead and tin are collected in the melting-pot."
— ***New Jerusalem Bible***

"Just as copper, tin, iron, and lead are melted down in a furnace."
— ***New Living Translation***

"As in collecting silver and copper and iron and lead and tin into the midst."
— ***New World Translation***

"[As] they gather silver, and copper, and iron, and lead, and tin, into the midst of the furnace."
— ***The Darby Translation***

"As they gather silver and copper and iron and lead and tin into the midst of a furnace."
— ***The Emphasized Bible***

No. 15
Ezekiel 27:12
(See exhibit C (commentary))

"For your wares, they exchanged silver, iron, tin, and lead."
 — *Common English Bible*

"Traded silver, iron, tin, and lead for your products."
 — *Contemporary English Version*

"With a multitude of all kinds of riches, with silver, iron, tin, and lead."
 — *Douay-Rheims*

"Your great wealth of every kind; silver, iron, tin, and lead they exchanged for your wares."
 — *English Standard Version*

"All kind of riches; with silver, iron, tin, and lead, they traded in thy fairs."
 — *King James Version*

"They traded your goods for silver, iron, tin, and lead."
 — *New Century Version*

"They exchanged silver, iron, tin and lead for your merchandise."
 — *New International Version*

"Trading your wares in exchange for silver, iron, tin, and lead."
 — *New Living Translation*

"The abundance of all substance; with silver, iron, tin, and lead."
 — *The Darby Translation*

"The multitude of all kinds of riches; with silver, iron, tin, and lead."
 — *World English Bible*

"The multitude of all kinds of riches, with silver, and iron, and tin, and lead."
 — *Wycliffe Bible*

"For silver, iron, tin, and lead, They have given out thy remnants."
 — *Young's Literal Translation*

No. 16
Matthew 3:4
(See exhibit F.)

"And his food was locusts and wild honey."
— ***Complete Jewish Bible***

"And his meat was locusts and wild honey."
— ***Douay-Rheims***

"His diet consisted of locusts and wild honey."
— ***God's Word Translation***

"His food was locusts and wild honey."
— ***Good News Bible***

"His food was locusts and wild honey."
— ***Hebrew Names Version***

"His food was locusts and wild honey."
— ***Lexham English Bible***

"His food was locusts and wild honey."
— ***New International Version***

"And his food was locusts and wild honey."
— ***New Jerusalem Bible***

"And his nourishment was locusts and wild honey."
— ***The Darby Translation***

"While, his food, was locusts and wild honey."
— ***The Emphasized Bible***

"And he lived upon locusts and wild honey."
— ***Weymouth New Testament***

"His nourishment was locusts and honey of the field."
— ***Young's Literal Translation***

No. 17
Romans 3:25
(See exhibit D.)

"Through faith, in his blood."
    — *American Standard Version*

"Through faith, by his blood."
    — *Bible in Basic English*

"Through faith in his blood."
    — *Douay-Rheims*

"Through faith in Christ's blood."
    — *God's Word Translation*

"Through faith, in his blood."
    — *Hebrew Names Version*

"Through faith in His blood."
    — *Holman Christian Standard*

"By his blood, to be received by faith."
    — *Revised Standard Version*

"Through faith in his blood."
    — *The Emphasized Bible*

"Through faith in his blood."
    — *The Webster Bible*

"Through faith in His blood."
    — *Weymouth New Testament*

"By faith in his blood."
    — *Wycliffe*

"Through the faith in his blood."
    — *Young's Literal Translation*

No. 18
2 Corinthians 6:6
(See exhibits B; B (continued))

"Knowledge and spiritual insight, longsuffering and patience."
— *Amplified Bible*

"In knowledge, in longsuffering, in sweetness."
— *Douay-Rheims*

"By our purity, knowledge, patience."
— *Good News Bible*

"In knowledge, in patience, in kindness."
— *Hebrew Names Version*

"By pureness, by knowledge, by long-suffering, by kindness."
— *King James Version, American Edition*

"In purity, in knowledge, in patience, in kindness."
— *Lexham English Bible*

"In purity, understanding, patience and kindness."
— *New International Version*

"Purity, our understanding, our patience, our kindness."
— *New Living Translation*

"By knowledge, by long-suffering."
— *New World Translation*

"In knowledge, in longsuffering, in kindness."
— *The Darby Translation*

"In knowledge, in long-suffering."
— *The Emphasized Bible*

"In knowledge, in long-suffering, in kindness."
— *Young's Literal Translation*

No. 19
Galatians 5:22–23
(See figure 3; exhibit A, B, C, D, E, and F.)

"Love, joy, peace, patience, kindness, goodness, faithfulness, humility, self control."
— ***Complete Jewish Bible***

"Loving, happy, peaceful, patient, kind, good, faithful, gentle, and self-controlled."
— ***Contemporary English Version***

"Love, joy, peace, patience, kindness, goodness, faithfulness, humility, and self-control."
— ***Good News Bible***

"Love, joy, peace, patience, kindness, goodness, faith, gentleness, self-control."
— ***Holman Christian Standard***

"Love, joy, peace, patience, kindness, goodness, faithfulness, gentleness, self control."
— ***Lexham English Bible***

"Love, joy, peace, forbearance, kindness, goodness, faithfulness, gentleness and self-control."
— ***New International Version***

"Love, joy, peace, patience, kindness, goodness, faithfulness, gentleness, and self-control."
— ***New Living Translation***

"Love, joy, peace, long-suffering, kindness, goodness, faith, mildness, self-control."
— ***New World Translation***

"Love, joy, peace, long-suffering, kindness, goodness, fidelity, meekness, self-control."
— ***The Darby Translation***

"Love, joy, peace, long-suffering, gentleness, goodness, faith, meekness, temperance."
— ***The Webster Bible***

"Love, joy, peace; patience ... kindness, benevolence; good faith, meekness, self-restraint."
— ***Weymouth New Testament***

"Love, joy, peace, patience, kindness, goodness, faith, gentleness, and self-control."
— ***World English Bible***

No. 20
Ephesians 6:14
(See exhibit A.)

"Take your place, then, having your body clothed with the true word, and having put on the breastplate of righteousness."
— *Bible in Basic English*

"Therefore, stand! Have the belt of truth buckled around your waist, put on righteousness for a breastplate."
— *Complete Jewish Bible*

"Let the truth be like a belt around your waist, and let God's justice protect you like armor."
— *Contemporary English Version*

"Fasten truth around your waist like a belt. Put on God's approval as your breastplate."
— *God's Word Translation*

"So stand ready, with truth as a belt tight around your waist, with righteousness as your breastplate."
— *Good News Bible*

"Stand therefore, having the utility belt of truth buckled around your waist, and having put on the breastplate of righteousness."
— *Hebrew Names Version*

"Stand therefore, girding your waist with truth, and putting on the breastplate of righteousness."
— *Lexham English Bible*

"Stand firm then, with the belt of truth buckled around your waist, with the breastplate of righteousness."
— *New International Version*

"Stand therefore, having girded your loins with truth, and having put on the breastplate of righteousness."
— *Revised Standard Version*

"Stand therefore, having girded your loins with truth, and put on the breastplate of righteousness."
— *The Emphasized Bible*

"Stand therefore, first fastening round you the girdle of truth and putting on the breastplate of uprightness."
— *Weymouth New Testament*

"Stand, therefore, having your loins girt about in truth, and having put on the breastplate of the righteousness."
— *Young's Literal Translation*

No. 21
Hebrews 10:38
(See exhibit A.)

"The people God accepts will live because of their faith."
— *Contemporary English Version*

"But the righteous will live by faith."
— *Hebrew Names Version*

"But My righteous one will live by faith."
— *Holman Christian Standard*

"But my righteous one will live by faith."
— *Lexham English Bible*

"But My righteous one shall live by faith."
— *New American Standard*

"But my righteous one will live by faith."
— *New International Version*

"My upright person will live through faith."
— *New Jerusalem Bible*

"And a righteous person will live by faith."
— *New Living Translation*

"But my righteous one shall live by faith."
— *Revised Standard Version*

"But, my righteous one, by faith, shall live."
— *The Emphasized Bible*

"But the righteous will live by faith."
— *World English Bible*

"And the righteous by faith shall live."
— *Young's Literal Translation*

No. 22
James 3:13
(See exhibit F.)

"His works in meekness of wisdom."
— *American Standard Version*

"Show that your actions are good with a humble lifestyle that comes from wisdom."
— *Common English Bible*

"The humility that grows out of wisdom."
— *Complete Jewish Bible*

"His work in the meekness of wisdom."
— *Douay-Rheims*

"Humility that comes from wisdom."
— *God's Word Translation*

"With humility and wisdom."
— *Good News Bible*

"His works with meekness of wisdom."
— *King James Version*

"Gentleness of wisdom."
— *New Jerusalem Bible*

"Mildness that belongs to wisdom."
— *New World Translation*

"His works with meekness of wisdom."
— *The Webster Bible*

"In mildness of his wisdom show he of good living his work in mildness of wisdom]"
— *Wycliffe*

"His works in meekness of wisdom."
— *Young's Literal Translation*

No. 23
1 Peter 3:10
(See exhibit C.)

"For it is said, Let the man who has a love of life."
— ***Bible in Basic English***

"For 'Whoever wants to love life and see good days.'"
— ***Complete Jewish Bible***

"For he that will love life and see good days."
— ***Douay-Rheims***

"For, 'He who would love life.'"
— ***Hebrew Names Version***

"A person must do these things to enjoy life."
— ***New Century Version***

"Whoever would love life and see good days."
— ***New International Version***

"If you want a happy life and good days."
— ***New Living Translation***

"For he that will love life and see good days."
— ***The Darby Translation***

"For, he that desireth to love, life, and to see good days."
— ***The Emphasized Bible***

"He who would love life, And see good days."
— ***World English Bible***

"For he that will love life, and see good days."
— ***Wycliffe***

"For 'he who is willing to love life.'"
— ***Young's Literal Translation***

No. 24
Revelation 9:9
(See exhibit A.)

"As it were breastplates of iron."
— *American Standard Version*

"And they had breastplates like iron."
— *Bible in Basic English*

"In front they had what seemed to be iron armor upon their chests."
— *Common English Bible*

"Their chests were like iron breastplates."
— *Complete Jewish Bible*

"And they had breastplates as breastplates of iron."
— *Douay-Rheims*

"With what looked like iron breastplates."
— *Good News Bible*

"They had breastplates, like breastplates of iron."
— *Hebrew Names Version*

"They had breastplates like breastplates of iron."
— *New International Version*

"They had breastplates as breastplates of iron."
— *The Darby Translation*

"They had breastplates as breastplates of iron."
— *The Emphasized Bible*

"They had breast-plates which seemed to be made of steel."
— *Weymouth New Testament*

"They had breastplates as breastplates of iron."
— *Young's Literal Translation*

List of all Bibles and corresponding verses cited.

1	American Standard Version
2	Amplified Bible
3	Bible in Basic English
4	Common English Bible
5	Complete Jewish Bible
6	Contemporary English Version
7	Douay-Rheims
8	English Standard Version
9	God's Word Translation
10	Good News Bible
11	Hebrew Names Version
12	Holman Christian Standard
13	King James Version
14	Lexham English Bible
15	New American Standard Bible
16	New Century Version
17	New International Version
18	New Jerusalem Bible
19	New Living Translation
20	New World Translation
21	Revised Standard Version
22	The Darby Translation
23	The Emphasized Bible
24	The Webster Bible
25	Weymouth New Testament
26	World English Bible
27	Wycliffe
28	Young's Literal Translation

American Standard Version
- 2   **Genesis 49:12**
- 3   **Exodus 15:10**
- 5   **Numbers 31:22**
- 8   **Job 13:14**
- 11   **Proverbs 8:10**
- 13   **Ezekiel 22:18**
- 17   **Romans 3:25**
- 22   **James 3:13**
- 24   **Revelation 9:9**

Amplified Bible
- 4   **Exodus 25:3**
- 6   **Deuteronomy 32:13–14**
- 7   **Job 6:12**
- 8   **Job 13:14**
- 9   **Psalms 38:3**
- 10   **Psalm 52:8**
- 11   **Proverbs 8:10**
- 13   **Ezekiel 22:18**
- 18   **2 Corinthians 6:6**

Bible in Basic English
- 1   **Genesis 2:7**
- 5   **Numbers 31:22**
- 7   **Job 6:12**
- 10   **Psalms 52:8**
- 17   **Romans 3:25**
- 20   **Ephesians 6:14**
- 23   **1 Peter 3:10**
- 24   **Revelation 9:9**

Common English Bible
- 3   **Exodus 15:10**
- 5   **Numbers 31:22**
- 9   **Psalms 38:3**
- 13   **Ezekiel 22:18**
- 14   **Ezekiel 22:20**
- 15   **Ezekiel 27:12**
- 22   **James 3:13**
- 24   **Revelation 9:9**

Complete Jewish Bible
- 1   **Genesis 2:7**
- 2   **Genesis 49:12**
- 3   **Exodus 15:10**
- 5   **Numbers 31:22**
- 6   **Deuteronomy 32:13–14**
- 7   **Job 6:12**

8	Job 13:14
10	Psalms 52:8
11	Proverbs 8:10
12	Proverbs 24:13–14
14	Ezekiel 22:20
16	Matthew 3:4
19	Galatians 5:22–23
20	Ephesians 6:14
22	James 3:13
23	1 Peter 3:10
24	Revelation 9:9

Contemporary English Version

2	Genesis 49:12
4	Exodus 25:3
6	Deuteronomy 32:13–14
13	Ezekiel 22:18
15	Ezekiel 27:12
19	Galatians 5:22–23
20	Ephesians 6:14
21	Hebrews 10:38

Douay-Rheims

1	Genesis 2:7
2	Genesis 49:12
5	Numbers 31:22
8	Job 13:14
9	Psalms 38:3
10	Psalms 52:8
12	Proverbs 24:13–14
15	Ezekiel 27:12
16	Matthew 3:4
17	Romans 3:25
18	2 Corinthians 6:6
22	James 3:13
23	1 Peter 3:10
24	Revelation 9:9

English Standard Version

4	Exodus 25:3
5	Numbers 31:22
6	Deuteronomy 32:13–14
8	Job 13:14
9	Psalms 38:3
11	Proverbs 8:10
12	Proverbs 24:13–14
15	Ezekiel 27:12

God's Word Translation

3	Exodus 15:10
4	Exodus 25:3
5	Numbers 31:22
13	Ezekiel 22:18
16	Matthew 3:4
17	Romans 3:25
20	Ephesians 6:14
22	James 3:13

Good News Bible

1	Genesis 2:7
3	Exodus 15:10
4	Exodus 25:3
6	Deuteronomy 32:13–14
7	Job 6:12
11	Proverbs 8:10
12	Proverbs 24:13–14
14	Ezekiel 22:20
16	Matthew 3:4
18	2 Corinthians 6:6
19	Galatians 5:22–23
20	Ephesians 6:14
22	James 3:13
24	Revelation 9:9

Hebrew Names Version

1	Genesis 2:7
2	Genesis 49:12
9	Psalms 38:3
10	Psalms 52:8
14	Ezekiel 22:20
16	Matthew 3:4
17	Romans 3:25
18	2 Corinthians 6:6
20	Ephesians 6:14
21	Hebrews 10:38
23	1 Peter 3:10
24	Revelation 9:9

Holman Christian Standard

3	Exodus 15:10
6	Deuteronomy 32:13–14
8	Job 13:14
13	Ezekiel 22:18
14	Ezekiel 22:20
17	Romans 3:25
19	Galatians 5:22–23
21	Hebrews 10:38

King James Version
- 4   Exodus 25:3
- 5   Numbers 31:22
- 7   Job 6:12
- 8   Job 13:14
- 9   Psalms 38:3
- 11   Proverbs 8:10
- 12   Proverbs 24:13-14
- 13   Ezekiel 22:18
- 15   Ezekiel 27:12
- 18   2 Corinthians 6:6
- 22   James 3:13

Lexham English Bible
- 16   Matthew 3:4
- 18   2 Corinthians 6:6
- 19   Galatians 5:22, 23
- 20   Ephesians 6:14
- 21   Hebrews 10:38

New American Standard Bible
- 4   Exodus 25:3
- 6   Deuteronomy 32:13–14
- 7   Job 6:12
- 8   Job 13:14
- 9   Psalms 38:3
- 10   Psalms 52:8
- 11   Proverbs 8:10
- 21   Hebrews 10:38

New Century Version
- 1   Genesis 2:7
- 2   Genesis 49:12
- 3   Exodus 15:10
- 6   Deuteronomy 32:13–14
- 11   Proverbs 8:10
- 13   Ezekiel 22:18
- 14   Ezekiel 22:20
- 15   Ezekiel 27:12
- 23   1 Peter 3:10

New International Version
- 1   Genesis 2:7
- 3   Exodus 15:10
- 4   Exodus 25:3
- 6   Deuteronomy 32:13–14
- 7   Job 6:12
- 11   Proverbs 8:10
- 12   Proverbs 24:13–14

14	Ezekiel 22:20
15	Ezekiel 27:12
16	Matthew 3:4
18	2 Corinthians 6:6
19	Galatians 5:22–23
20	Ephesians 6:14
21	Hebrews 10:38
23	1 Peter 3:10
24	Revelation 9:9

New Jerusalem Bible

3	Exodus 15:10
4	Exodus 25:3
7	Job 6:12
10	Psalms 52:8
11	Proverbs 8:10
12	Proverbs 24:13–14
14	Ezekiel 22:20
16	Matthew 3:4
21	Hebrews 10:38
22	James 3:13

New Living Translation

3	Exodus 15:10
7	Job 6:12
14	Ezekiel 22:20
15	Ezekiel 27:12
18	2 Corinthians 6:6
19	Galatians 5:22–23
21	Hebrews 10:38
23	1 Peter 3:10

New World Translation

2	Genesis 49:12
4	Exodus 25:3
5	Numbers 31:22
7	Job 6:12
9	Psalms 38:3
10	Psalms 52:8
12	Proverbs 24:13–14
14	Ezekiel 22:20
18	2 Corinthians 6:6
19	Galatians 5:22–23
22	James 3:13

Revised Standard Version

1	Genesis 2:7
2	Genesis 49:12
5	Numbers 31:22

       9   **Psalms 38:3**
     13   **Ezekiel 22:18**
     17   **Romans 3:25**
     20   **Ephesians 6:14**
     21   **Hebrews 10:38**

The Darby Translation

      1   **Genesis 2:7**
      2   **Genesis 49:12**
      4   **Exodus 25:3**
      5   **Numbers 31:22**
      7   **Job 6:12**
      8   **Job 13:14**
      9   **Psalms 38:3**
     11   **Proverbs 8:10**
     12   **Proverbs 24:13–14**
     13   **Ezekiel 22:18**
     14   **Ezekiel 22:20**
     15   **Ezekiel 27:12**
     16   **Matthew 3:4**
     18   **2 Corinthians 6:6**
     19   **Galatians 5:22–23**
     23   **1 Peter 3:10**
     24   **Revelation 9:9**

The Emphasized Bible

      1   **Genesis 2:7**
      2   **Genesis 49:12**
      4   **Exodus 25:3**
      6   **Deuteronomy 32:13–14**
      7   **Job 6:12**
      8   **Job 13:14**
     10   **Psalms 52:8**
     12   **Proverbs 24:13–14**
     14   **Ezekiel 22:20**
     16   **Matthew 3:4**
     17   **Romans 3:25**
     18   **2 Corinthians 6:6**
     20   **Ephesians 6:14**
     21   **Hebrews 10:38**
     23   **1 Peter 3:10**
     24   **Revelation 9:9**

The Webster Bible

      1   **Genesis 2:7**
      3   **Exodus 15:10**
      6   **Deuteronomy 32:13–14**
     10   **Psalms 52:8**
     13   **Ezekiel 22:18**

17	Romans 3:25
19	Galatians 5:22–23
22	James 3:13

Weymouth New Testament
16	Matthew 3:4
17	Romans 3:25
19	Galatians 5:22–23
20	Ephesians 6:14
24	Revelation 9:9

World English Bible
6	Deuteronomy 32:13–14
10	Psalms 52:8
13	Ezekiel 22:18
15	Ezekiel 27:12
19	Galatians 5:22–23
21	Hebrews 10:38
23	1 Peter 3:10

Wycliffe
1	Genesis 2:7
2	Genesis 49:12
3	Exodus 15:10
8	Job 13:14
9	Psalms 38:3
12	Proverbs 24:13–14
15	Ezekiel 27:12
17	Romans 3:25
22	James 3:13
23	1 Peter 3:10

Young's Literal Translation
2	Genesis 49:12
5	Numbers 31:22
8	Job 13:14
9	Psalms 38:3
10	Psalms 52:8
11	Proverbs 8:10
12	Proverbs 24:13–14
15	Ezekiel 27:12
16	Matthew 3:4
17	Romans 3:25
18	2 Corinthians 6:6
20	Ephesians 6:14
21	Hebrews 10:38
22	James 3:13
23	1 Peter 3:10
24	Revelation 9:9

Some of the Bible verses were from the following online locations:

        http://www.blueletterbible.org/search.cfm
        http://www.biblegateway.com/
        http://www.catholic.org/bible/
        http://www.o-bible.com/bbe.html
        http://www.biblestudytools.com/cjb/
        http://rockhay.tripod.com/worship/translat.htm
        http://lookhigher.net/englishbibles/theemphasisedbible/genesis/1.html
        http://www.newlivingtranslation.com/default.asp
        http://www.mystudybible.com/

vjoneal@yahoo.com

22000Z00420402Z055231

www.ingramcontent.com/pod-product-compliance
Lightning Source LLC
Chambersburg PA
CBHW051025180526
45172CB00002B/476